Gravity

A Universal Tug-of-War

Now I Get It!

SAN DIEGO PUBLIC LIBRARY
LA JOLLA

3 1336 10921 6409

Glen Phelan

Sally Ride, Ph.D., President and Chief Executive Officer;
Tam O'Shaughnessy, Chief Operating Officer and
Executive Vice President; Margaret King, Editor;
Monnee Tong, Design and Picture Editor; Erin Hunter,
Science Illustrator; Brenda Wilson, Editorial Consultant;
Matt McArdle, Editorial Researcher

Program Developer, Kate Boehm Jerome
Program Design, Steve Curtis Design Inc.
www.SCDchicago.com

Copyright © 2011 by Sally Ride Science. All rights
reserved. No part of this book may be reproduced or
transmitted in any form or by any means, electronic or
mechanical, including photocopying, recording, or by
any information storage and retrieval system, without
permission in writing from the publisher.

Sally Ride Science is a trademark of Imaginary Lines, Inc.

Sally Ride Science
9191 Towne Centre Drive
Suite L101
San Diego, CA 92122

ISBN: 978-1-933798-51-6

Printed in the United States of America
10 9 8 7 6 5 4 3 2 1
First Edition

Cover: Skydivers fall through the air because gravity
pulls them toward Earth.

Title page: A chunk of ice breaks off and crashes into
the sea as gravity tugs it down.

Right: Gravity pulls the comet Hale-Bopp, seen here in
1997, into a long orbit around the Sun.

*Sally Ride Science is committed to minimizing its environmental impact by using
ecologically sound practices. Let's all do our part to create a healthier planet.*

*This book is printed on paper made with 100% recycled fiber, 50% post-consumer
waste, bleached without chlorine, and manufactured using 100% renewable energy.*

Contents

Introduction

In Your World

Wicked twists, super grabs, giant flips . . . just another incredible run at the snowboarding championships.

These athletes use skill, strength, and flexibility to make their world-class moves. They need something else, too. Without it, they wouldn't get out of the starting gate. Can you guess what it is? It's the same thing that brings the falling rain, puts the fun into a cannonball dive, and sends chunks of loose ice tumbling downhill. It's **gravity**!

Gravity isn't something you can see, but you sure can see what it does. It pulls things toward Earth. It pulls little things like raindrops, big things like boulders, and medium-sized things like . . . snowboarders.

That's not all. Gravity does much more. It keeps the Moon circling Earth and Earth circling the Sun. Gravity even causes stars to form!

How can something that makes your pencil roll off the desk have such awesome, far-reaching power? Read on!

What Goes Up . . .

Can you finish the title of this chapter? What goes up . . . must come down! This old saying could describe a fly ball hit to the outfield, a gymnast in the middle of a tumble run, or anything else that rises into the air. In short, the saying describes the effects of gravity—at least the way gravity works here on Earth.

Gravity is a **force** that pulls together any two objects—like the gymnast and Earth. No matter how high the gymnast launches herself into the air, she falls back to the ground. That's because Earth's gravity pulls her down. That is, it pulls her toward Earth. Gravity tugs on a dead branch that breaks off a tree and those shoes you toss that flop onto the floor. They fall because Earth's gravity pulls on them.

Earth's gravity pulls on you, too. That's a good thing. Without gravity, you'd float right off the planet into space!

▼ **Gravity keeps this gymnast, and everything on Earth, grounded.**

Stranger Than Fiction

Speaking of space, gravity works there, too. In fact, gravity is universal—it acts between any two objects in the Universe. Think about what that means!

Name any two objects—big or small, indoors or outdoors. It doesn't matter what or where they are. Gravity between those objects is pulling them together!

If gravity is pulling every pair of objects together, why isn't everything in the Universe clumped together into one big blob? Why doesn't this book "fall" toward you?

The answer is that gravity is a very, very, *very* weak force. Gravity really *is* pulling you and this book together, but the pull is too small to notice. It's not like **magnetism**. That's a much stronger force! If you slowly roll a small metal ball past a magnet, you actually can see the ball being pulled toward the magnet.

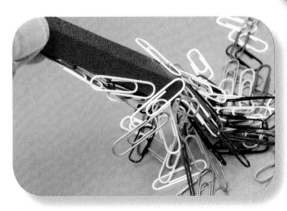

▲ Magnetism is a much stronger force than gravity.

Never Pushy

Magnetism can push or pull. Magnetic poles that are alike push away, or repel, each other. Gravity never pushes—ever. If gravity in your classroom suddenly pushed instead of pulled, everything that wasn't anchored would be plastered against the ceiling!

The Bottom Line | Gravity is a very weak force that pulls any two objects together.

7

Tug-of-war Champion

You might be wondering—if gravity is so weak, how can it make all objects on Earth fall down? Here's part of the answer. The strength of gravity between two objects depends on the **mass** of the objects. Mass is the amount of **matter** something has. The greater the mass, the stronger the pull.

Earth has a tremendous mass, so its gravitational pull is tremendous. By comparison, the gravitational attraction from other things around you on Earth isn't noticeable. If you drop your backpack, it falls to the ground, not toward you. Your body is pulling on it, but so is Earth. That's a game of tug-of-war that Earth wins hands down!

▼ **The trees and rocks pull on the water, so why doesn't the water fall toward them?**

To the Moon and Beyond!

To get a better idea of how mass affects gravity, let's go to the Moon. The Moon has about one-eightieth (1/80) the mass of Earth. That's still fairly massive, so the Moon pulls objects toward it. However, objects on the Moon fall only about one-sixth (1/6) as fast as on Earth, sort of like a rock falling through water.

The Moon may not be as massive as Earth, but the Sun is a different story. It has more than 330,000 times the mass of our planet!

If the Sun is so massive, why does a tossed coin fall to the ground instead of fly toward the Sun? Why doesn't the more massive object—the Sun—pull the coin toward it? It's because the strength of gravity between two objects depends on the distance between them as well as their masses. The farther apart they are, the weaker their gravitational pull on each other. The Sun is 150 million kilometers (93 million miles) away. That's too far to overcome Earth's pull on the coin.

▲ With one-sixth (1/6) the gravity of Earth, the Moon is a fun place to take a jump!

The Wow!

Monster Dunks

Want to dominate the basketball league? Go to the Moon! With only one-sixth (1/6) of Earth's gravity, you could jump six times higher than on Earth. Think of the monster dunks you could make! Of course, the other team could also jump six times higher and block you.

The Bottom Line | The strength of gravity between two objects depends on two things—their masses and the distance between them.

Look Out Below!

By now you probably have a good idea of how Earth's gravity affects your everyday life. For one thing, it keeps you anchored to our planet!

Earth's gravity constantly pulls you toward the center of the planet. Of course, things like floors and the ground block your way, but Earth is still pulling you toward its center. Don't believe it? Then think about jumping into a hole in the ground. If Earth's pulling force stopped at ground level, there'd be nothing to make you fall into the hole! You'd hover over it or fall back toward the ground, instead of falling in.

▼ **The effects of gravity in your classroom are not noticeably different from what they are at the top of the world—Mount Everest.**

Faster, Faster!

Hold up a small object—unbreakable, please—and drop it. What happens to it on the way down? Because of Earth's gravity, there's more to a fall than meets the eye.

A high-speed camera photographed a dropped basketball every one-twentieth (1/20) of a second. This drawing shows what happened. Notice that the ball falls a greater and greater distance every one-twentieth of a second. That means it moves faster and faster as it falls. In other words, it **accelerates**.

Gravity makes the ball accelerate, or speed up, while it's falling. Every falling object accelerates, whether it's a rock that breaks off a ledge or a pencil that drops off your desk. The really cool thing is that all objects accelerate at the same rate!

All objects falling near Earth's surface have an acceleration of 9.8 meters per second squared. This means the **velocity** of a falling object increases by 9.8 meters per second (9.8 m/s) every second that it falls. So the ball, the rock, and your pencil all fall 9.8 m/s after one second, 19.6 m/s (9.8 m/s + 9.8 m/s) after two seconds, and so on. But if you think *that's* cool, just wait!

Falling Objects Accelerate

◄ **This drawing shows the position of a basketball every one-twentieth (1/20) of a second as it falls. It falls from the top position to the bottom position in one-half (1/2) second. The ball gains speed as it falls. How fast would it be falling after three seconds?**

The Bottom Line | Earth's gravity makes all falling objects accelerate at the same rate—9.8 m/s².

It's a Tie!

Suppose a diver holds a ball but drops it as she begins her dive. Which will hit the water first—the ball or the diver? Both objects start falling from the same height at the same time, but the diver has a much greater mass than the ball. Since more mass means more gravity, the diver should hit the water first, right?

Not quite. It's true that the force of gravity is greater between Earth and the larger mass. But it's also true that more force is needed to change the motion of the object with the larger mass. That's why you have to pull harder to move a loaded wagon than an empty one.

So, the force on the diver is greater than on the ball, but it's harder to move the diver. The net result is that the diver and the ball fall at the same rate—9.8 m/s².

▲ The diver *accelerates* at the same rate as a ball, a marble, or an elephant!

Oh, and in case you haven't guessed it yet, the ball and the diver hit the water at the same time. How can this be? Check out the formulas.

▼ As the amount of mass increases so does the amount of force needed to move it—so the ratio of f/m, or acceleration, stays the same!

Diver		Ball	
Acceleration =	$\dfrac{force\ of\ gravity}{mass\ of\ diver}$	Acceleration =	$\dfrac{force\ of\ gravity}{mass\ of\ ball}$
Acceleration =	$\dfrac{539\ newtons}{55\ kilograms}$	Acceleration =	$\dfrac{19.6\ newtons}{2\ kilograms}$
Acceleration =	$\dfrac{9.8\ m}{s^2}$	Acceleration =	$\dfrac{9.8\ m}{s^2}$

You Must Resist

You still may have doubts that everything falls at the same rate. After all, you *know* some things fall more slowly than others. A flat sheet of paper, for instance, falls more slowly than the same sheet crumpled into a ball. So what's going on?

Air resistance is what's going on! Air resistance is the force of air molecules pushing against a moving object. Air resistance pushes up on a falling object and slows it down. Air resistance affects different objects differently.

A flat sheet of paper has a larger surface area than a crumpled sheet. The larger surface area comes in contact with more air molecules and increases air resistance. This slows the paper's fall. The flat sheet of paper flutters to the ground, but the crumpled paper falls like a rock.

The Wow!

Here's Proof

If you drop a feather and a hammer at the same time, the hammer hits the ground first. Air resistance slows the feather. But what happens if you drop these objects on the Moon? An astronaut tried it, and they hit the ground at the same time! There was no air resistance because there's no air on the Moon!

◀ Without air resistance, parachutes wouldn't work.

The Bottom Line

Objects fall at the same rate if gravity is the only force acting on them. But air resistance also acts on objects and affects how fast they fall.

A Weighty Topic

Picture this. You and an elephant are floating in space. Too weird? Fine, make it a hippopotamus. So, you and this hippo are floating in space far away from Earth and everything else. You remember that astronauts and other things in space are weightless—they don't weigh anything. So, you figure it's the perfect time to play "Push the Hefty Hippo." You drift over and give your fellow space traveler a shove. To your surprise, the hippo barely budges! "Why not?" you wonder. "Why can't I push something that weighs practically nothing?" The answer is that you're confusing **weight** with mass.

▼ **Why can astronauts float in space?**

Weight is the force of gravity on an object. When you step on a bathroom scale to weigh yourself, you're measuring how much Earth's gravity is pulling you down onto the scale. The more mass you have, the more gravity pulls and the more you weigh.

Mass Stays the Same

Out in deep space, you and the hippo are so far from Earth—or any other large object—that the pull from Earth's gravity is almost zilch. So, you weigh almost zilch. In fact, if you floated over to a scale and tried to stand on it, you wouldn't be able to. There wouldn't be enough gravity to pull you toward it!

In space, your weight has changed, but your mass hasn't. Mass is how much matter something has. Weight is how much gravity pulls on that matter.

As for that weightless hippo, you couldn't move it much because its mass hasn't changed in space. It's still a huge hippo!

The mass of an astronaut on Earth and on the Moon is the same. If the weight of an astronaut is 72.5 kilograms on Earth, what is the weight of the astronaut on the Moon? Why?

The Wow!

Planetary Weight Plan

How can you lose or gain weight without changing the way you eat? Go to Mars . . . or Jupiter—or any other large body in the **solar system**. Your weight depends on the planet's mass and how far you are from its center. On Mars, you'd weigh about one-third of your weight on Earth. On Jupiter, you'd weigh two and a half times as much!

The Bottom Line | Weight and mass are not the same. Mass is how much matter something has, and weight is how much gravity pulls on matter.

Keeping It Together

Did you ever gaze up at a starlit sky? It's an awesome sight! The stars look like tiny specks of glitter scattered across the sky. But looks can be deceiving—stars are the largest objects in the Universe.

Stars are deceiving in another way. Their locations may seem random, but these gigantic balls of hot, glowing gases are actually quite organized. All the stars you can see without a **telescope**, including our own Sun, are part of a large group of stars called the Milky Way **galaxy**. The galaxy is held together by a certain force.

What is this great universal organizer? You've guessed it—gravity!

Remember, even though gravity is a weak force, it packs a powerful pull when objects are massive. Well, objects don't get much more massive than stars! Bunches of stars are held together by gravity in galaxies throughout the Universe. Planets aren't exactly lightweights, either. Within galaxies, gravity holds together stars and their planets to form different solar systems. Let's see how gravity formed our own solar system!

▼ **Gravity keeps all these stars together in the same group in space. Still, most of them are billions of kilometers apart.**

▲ Gravity pulled enough gas together to form our Sun. How do you think the planets formed?

Building with Gravity

Long, long ago, the little part of the Universe where we live was a very different place. There was no Sun, no planets, no moons—just a huge, slowly spinning cloud of gas and dust.

Then, about 4.6 billion years ago, you might say the neighborhood began to change. Gravity started pulling the gas and dust together. As the cloud shrank, it spun faster. The spinning cloud flattened into a disk—the way pizza dough flattens as it spins while tossed in the air.

Gravity continued to pull material into a tighter and tighter ball in the center of the disk. As the ball became denser, it heated up.

Gravity continued to compress the ball of gas. It became so dense and hot that its atoms collided and combined. As the atoms combined, they released lots of energy. After a few million years, the force of the energy pushing out balanced the force of gravity pulling in, and the ball stopped compressing. A star—our Sun—was born!

The Bottom Line

The pull of gravity acts over very long distances.

Caution! Construction Ahead

The Sun was a good start, but the formation of our solar system was far from over. The Sun's gravity kept material in the disk moving around, like a giant merry-go-round. Farther from the Sun's intense heat, the gas in the disk cooled and formed bits of metal and rock. These particles collided over millions of years and stuck together, forming clumps the size of boulders.

Wham! Bam! Countless collisions and millions of years later, the boulders grew large enough to develop a sizable amount of gravity. This gravity swept up the remaining nearby material. The boulders grew larger and larger until they became massive enough to pull themselves into the shape of balls. These balls of material grew until they became large enough to be planets. After about 100 million years, the neighborhood was finally taking shape!

So THAT's Why!

Some large, round objects in our solar system never quite made it to planet size. They're called dwarf planets. The largest dwarf planet is Eris. It's so big it even has a moon. You might have heard of another icy dwarf planet, Pluto. Poor Pluto. It used to be called a planet. But once other ice balls were discovered, astronomers downgraded Pluto to a dwarf.

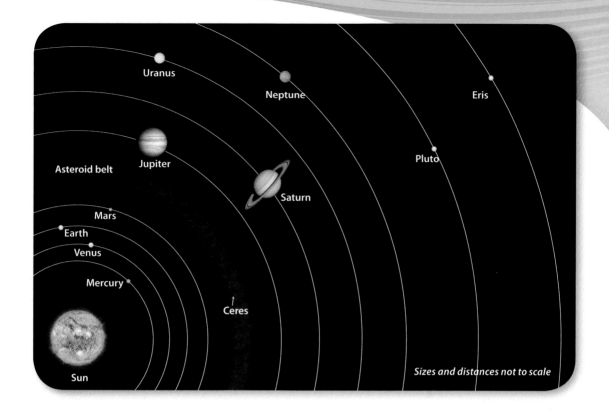

Uranus

Neptune

Eris

Jupiter

Pluto

Asteroid belt

Saturn

Mars

Earth

Venus

Mercury

Ceres

Sun

Sizes and distances not to scale

Welcome to the Neighborhood

So, let's meet some of the neighbors! They include eight planets, at least five dwarf planets, thousands of smaller rocky **asteroids**, and millions of icy rocks called **comets**. In the middle of it all is our Sun. The force of gravity between the Sun and the other objects holds them in predictable **orbits** around the Sun.

Some of our neighbors take a long, long time to complete an orbit because they are so far away. Yet gravity holds them in place. It takes Earth one year, or 365 days, to orbit the Sun. Mars is farther away from the Sun than Earth is and takes about twice as long to orbit the Sun. Eris, a dwarf planet in the outer region of our solar system, takes 557 years!

The Bottom Line

The Sun's gravity holds together all the planets, dwarf planets, asteroids, and comets in our solar system.

"Now Pitching . . ."

How does gravity keep objects like planets and satellites in orbit? To answer that question, think about throwing a baseball from the top of a very tall mountain.

What happens if you throw the ball straight ahead? The force of your throw pushes the ball forward. At the same time, gravity pulls it down. The combination of your throw and gravity makes the ball fall in an arc. The faster you throw the ball, the farther it will travel and the longer the arc will be.

Now, suppose you have an arm like Superwoman and throw the ball with a velocity of about 8,000 meters per second, which is about 17,500 miles per hour! The arc of the fall would match the curve of Earth. The ball would keep falling, but it would never hit the ground. That's because it would actually be falling *around* Earth, not into it. The ball would be in Earth's orbit!

Throw a ball faster and faster, and it falls farther and farther away in an arc. At a speed of 8,000 m/s the ball would be traveling so fast that the arc would match Earth's curve. The ball would literally fall *around* Earth and never touch the ground!

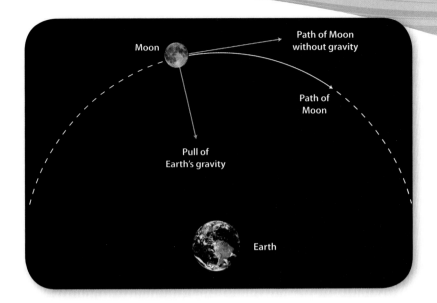

Moon

Path of Moon without gravity

Path of Moon

Pull of Earth's gravity

Earth

◀ Gravity pulls the Moon into orbit around Earth the way you might pull a tethered ball into a circle around you.

Round and Round

Once the ball is in orbit, gravity and **inertia** keep it there. Inertia just means an object continues to do what it's doing until a force acts on it.

Inertia enables the orbiting baseball to travel with the velocity you gave it, straight ahead at 8,000 meters per second. But wait a minute—it's not moving straight ahead. It's moving in a big circle. Aha! That's because a force—gravity—is acting on the ball. Without gravity, the ball's inertia would make it fly straight out into space. Without inertia, gravity would pull the ball straight down to Earth. The combination of inertia and gravity keeps the ball falling in a circular path around Earth.

So THAT's Why!

You've seen pictures of astronauts inside their orbiting space station. Why does everything float? It's because the space station and everything in it are falling around Earth at the same rate. If the astronauts had a scale with them, it would be falling, too. They wouldn't be able to stand on the scale to register a weight. So you might say the astronauts are weightless.

The Bottom Line

Gravity holds objects in orbit.

▲ The Hubble Space Telescope in Earth's orbit peered deep into space to take this photo. Every patch of light is a galaxy.

In a Galaxy Far, Far Away . . .

You've seen how gravity organizes our solar system. Now, take a look at what it does in deep space! Gravity holds stars and other objects together in massive groups called galaxies. How massive? Each galaxy contains more than 10 million stars. Some are home to almost a trillion stars!

But gravity's work doesn't stop with holding together stars in groups. Galaxies aren't just scattered about the Universe—gravity pulls them together, too, into clusters. A cluster might contain just a few galaxies, or it might contain a thousand of them. The gravity from many clusters forms an even larger group. What do you think they're called? Superclusters, what else?

A Powerful Force

Our Sun is just one of about 100 billion stars in the Milky Way. All of these stars orbit the galaxy's center like planets around the Sun. Not only does Earth travel around the Sun every year, our entire solar system is on a long journey around the Milky Way's center. It's trillions and trillions of kilometers away! How long is the journey? About 200 million years! And we're not exactly moving at a snail's pace. The solar system zooms through space at 240 kilometers (150 miles) per second.

Such distances and speeds are mind-boggling. So is this—astronomers have discovered that a supermassive **black hole** lies at the center of the Milky Way. How massive is it? It's as massive as four *million* Suns. Now *that* has a huge gravitational pull.

Black holes and the full effect of their gravity are still being studied. But one thing is certain—gravity governs motion throughout the Universe. Think about *that* the next time your pencil rolls off your desk!

Our Solar System

▲ The Milky Way is shaped like a pinwheel. Our solar system is located on one of its spiral arms.

The Wow!

Star Show

Sometimes gravity puts on quite a show. As a star grows old, it no longer produces enough energy at its center to hold off the force of gravity. The star begins to collapse. Temperature and pressure build and build until the star explodes in a supernova like Cassiopeia A, above.

The Bottom Line

A galaxy is a group of many stars held together by gravity. Our solar system is part of the Milky Way galaxy.

THINKING LIKE A
SCIENTIST

Exploring other worlds, up close and personal, is one of the most exciting things scientists do. But how do they do it? What do they have to think about in their plans? Well, gravity is one thing they must consider.

▶ A rocket must get moving very fast, very quickly to escape Earth's gravity.

Investigating

Think about a robotic space mission to Mars. The work and planning start years before the spacecraft leaves Earth. There are many problems to solve. Scientists figure them out in an organized way.

▼ Landing a craft on Mars requires hitting a moving target.

Problem Solving

Many of the problems scientists investigate involve gravity. To escape Earth's gravity, a rocket must travel at least 11,000 meters (36,089 feet) per second. Scientists and engineers have to consider the total weight of the rocket to make sure it can reach this speed.

Reaching Mars means the scientists have to hit a moving target. They have to know exactly when Mars is going to be at a certain place in its orbit. Mission planners might even use the Moon's gravity to save fuel and slingshot the spacecraft toward its destination.

Scientists also need to know the strength of gravity on Mars. Why? Remember, the force of gravity is different on different planets. An object will weigh either more or less than it does on Earth.

Interpreting Data

Scientists know that a good way to compare weights is to use a table. The one here shows numbers that you can use to multiply by and find an object's weight throughout our solar system.

Your turn! Use the information in the table to answer these questions.

Body in Space	Gravitation Factor Compared to Earth (Earth = 1)
Sun	27.90
Mercury	0.38
Venus	0.91
Moon	0.17
Mars	0.38
Jupiter	2.54
Saturn	1.08
Uranus	0.91
Neptune	1.19
Pluto	0.06

1. A spacecraft weighs 349 kilograms (770 pounds) on Earth. What will it weigh on Mars?

2. What would the spacecraft weigh on Jupiter?

3. The legs of the spacecraft are designed to support up to 544 kilograms (1,200 pounds). Could it be used on Venus? Explain.

4. How might a landing craft be designed differently for Pluto than for Mars? Why might scientists want to design it differently for Pluto?

▼ **The legs of a landing craft may not have to be as large or strong to support the craft on Mars as they would on Earth.**

Gravity's Spotlight on the Universe

Gravity is the cosmic glue that holds together the Universe. Gravity pulls together material, or matter, to form shining stars. It draws together stars to form galaxies, and pulls together galaxies in larger groups called clusters. Gravity gives the Universe order.

Some matter in the Universe, like stars, is visible—it gives off light. Other matter, like planets and moons, can only be seen because light bounces off it. But surprisingly, most matter is "dark"—it doesn't give off or reflect light. It's invisible.

How do we know it's there?

Even if astronomers can't see dark matter, it affects things they can see—like starlight. How? All matter, visible or invisible, has mass. And all mass has gravity. A cluster of galaxies might contain so much dark matter that its gravity bends the path that starlight travels through space. Yes, gravity bends light! When astronomers see starlight being bent, it's like they have stumbled across a sign that says, "Lots of mass—and gravity—here."

In this way, gravity helps astronomers understand where and how matter is arranged in space, even if that matter is invisible. In a way, gravity shines a light on the Universe's structure.

▶ Four galaxy clusters collided to form MACS J0717. It is one of the most complex galaxy clusters ever spotted.

Dara Norman

Astronomer

NATIONAL
OPTICAL
ASTRONOMY
OBSERVATORY

NOAO

◀ Dara's travels took her to Iguazu Falls, Argentina.

Dara Norman's ticket to a career in science only cost a handful of dimes. Growing up, Dara spent hours roaming Chicago's Museum of Science and Industry. "Back then, the museum was free. It only cost me bus fare," Dara remembers.

The experience inspired young Dara to study space. "I've wanted to be an astronaut from third grade on," she says. So, Dara would get her mom's permission to be late for school to watch Space Shuttle launches. Later, Dara began exploring space on her own, using

▲ Dara celebrates her ninth birthday.

telescopes in college and graduate school. What an eye-opener! "I was a real city kid—I had never really seen the Milky Way before."

Today, Dara does research at **observatories** around the world. "Traveling to places like Chile or the Canary Islands is always a big plus!" Telescopes are just one tool Dara uses to study quasars, the bright centers of active galaxies. Another tool is gravity, which can brighten a quasar's light. "Since astronomers work with light, and gravity affects light, it's very important in our work," Dara says.

▼ Dara's at home inside the telescope dome. The staircase climbs the structure that supports the telescope.

Dara Norman steps outside the dome, above, shielding a telescope high in the foothills of the Andes Mountains of Chile. It's a cold night at the observatory. Perfect—no clouds. That gives Dara many more hours to use the telescope. She quickly gets back to snapping digital pictures of thousands of faint galaxies. Some images capture the glint of brighter but more distant galaxies. Dara knows the galaxies in the foreground are massive. Their gravity can bend and even brighten the passing light from the more far-off galaxies. Dara wonders about gravity's effect. Is she seeing more of the faraway galaxies only because gravity brightens their light? If so, maybe those bright lights pinpoint the places where mass is concentrated in the Universe. Maybe they point to where mysterious dark matter could be found.

The Hubble Space Telescope, below, captured an image, right, of galaxy cluster Abell 1689. It's one of the most massive objects in the Universe. The mass of the cluster's trillion stars—plus its dark matter—bends the passing light of more distant galaxies. That makes the galaxies appear more curved— as if seen through an enormous lens. The effect is called gravitational lensing. It's a natural phenomenon and another tool astronomers use to do their job.

▶ This image shows the fuzzy yellow galaxies within Abell and a few fuzzy blue distant galaxies.

MATH CONNECTION

Weigh Different!

How do you measure the gravitational pull between Earth and you? Step on a scale! Scales measure weight—the force of attraction between you and Earth. The amount of attraction depends on the mass of each object and how far apart they are. Find out how much you weigh on some other bodies in the Universe. Do the math.

Place	Gravitational Pull
Sun	28 times that of Earth
Moon	$\frac{1}{6}$ that of Earth
Mars	$\frac{1}{3}$ that of Earth
White Dwarf	1.3 million times that of Earth
Neutron Star	140 billion times that of Earth

Oh, By the Way

Dara uses eyeglasses to explain gravitational lensing. "My eyes cannot focus the light coming into them properly. So I wear glasses. The lenses bend the light in the right way and focus it on my retinas. Gravity acts like a lens in the same way— it bends the light."

Hey, I Know THat!

From dazzling snowboarders to exploding stars—you've learned a lot about how gravity affects the Universe and your little part of it. On a sheet of paper, show what you know as you do the activities and answer these questions.

1. What two things affect the strength of gravity between two objects? (pages 8 and 9)

2. A softball player hits a high fly ball. On its way down, the softball falls 9.8 m/s for the first second. What is its velocity after 5 seconds? (page 11)

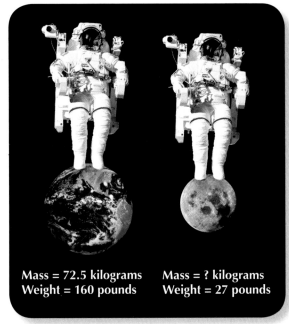

Mass = 72.5 kilograms
Weight = 160 pounds

Mass = ? kilograms
Weight = 27 pounds

3. Look at the photo of the astronaut on Earth and the Moon. Why is the astronaut's weight less on the Moon than on Earth? (pages 14 and 15)

4. What is the mass of the astronaut on the Moon? (page 15)

5. Draw a sketch to show how inertia and gravity combine to keep the Moon or a spacecraft in orbit around Earth. (pages 20 and 21)

Glossary

acceleration (n.) the rate at which an object's velocity changes—speeds up, slows down, or changes direction—with time (p. 11)

air resistance (n.) friction, or drag, that acts on something moving through air (p. 13)

asteroid (n.) a small rocky object that orbits the Sun. Thousands of asteroids orbit in a region called the Asteroid Belt, which lies between the orbits of Mars and Jupiter. However, some have been found in other orbits, including some that cross Earth's orbit. (p. 19)

black hole (n.) a concentration of mass near which gravity is so intense that not even light can escape (p. 23)

comet (n.) a mass of ice, rock, and gases that orbits the Sun (p. 19)

force (n.) a push or pull applied to an object (p. 6)

galaxy (n.) a large collection of stars bound together by gravity. Our Sun is one of many stars in the Milky Way galaxy. (p. 16)

gravity (n.) the attractive force that any object with mass has on all other objects with mass. The greater the mass of the object, the stronger its gravitational pull. (p. 5)

inertia (n.) the tendency of an object to keep doing what it's doing and resist any change in its state of motion. Mass is the measure of inertia. (p. 21)

magnetism (n.) the ability to attract objects made of iron (p. 7)

mass (n.) a measure of the total amount of matter contained within an object (p. 8)

matter (n.) any substance that has mass and takes up space (p. 8)

observatory (n.) a building designed to hold telescopes, computers, and other equipment for astronomers to use for observing planets, stars, and other celestial objects. Most observatories are dome shaped and able to rotate so telescopes can look in any direction. (p. 27)

orbit (n.) the path of one body around another as a result of the force of gravity between them. Examples are a planet's path around the Sun or a moon's path around a planet. (p. 19)

solar system (n.) a star and all objects, such as the planets, that revolve around it (p. 15)

telescope (n.) an instrument used to collect and focus light to produce a magnified image of a faraway object (p. 16)

velocity (n.) the speed and direction of an object's motion (p. 11)

weight (n.) a measure of the force of gravity on an object, or weight = mass x acceleration of gravity (p. 14)

Index

About the Author Glen Phelan has shared his fascination with science through teaching and writing—a fascination sparked as a teenager by the lunar missions of the Apollo Program. Learn more at www.sallyridescience.com.

Photo Credits Joggie Botma: Cover. © Borut Trdina: Back cover. Kerry Banazek: Title page. © Kenneth C. Zirkel: p. 2. Ljupco Smokovski: p. 4. Associated Press: p. 6. Thomas Mounsey: p. 7. Mary Terriberry: p. 8. NASA/JSC: p. 9 top. REUTERS/Stringer: p. 10. Lawrence Sawyer: p. 12. Kenneth William Caleno: p. 13 left. NASA Goddard Space Flight Center: p. 13 right. NASA: p. 14, p. 15 (Earth), p. 25. NASA/JPL: p. 15 (Moon). NASA, ESA, and the Hubble Heritage Team (AURA/STScI): p. 15 (Jupiter). Jimmy Westlake (Colorado Mountain College): p. 16. NASA/JPL-Caltech: p. 17. Robert Hurt (IPAC): p. 18. Image created by Reto Stockli with the help of Alan Nelson, under the leadership of Fritz Hasler: p. 20, 21 (Earth in illustration). Robert Williams and the Hubble Deep Field Team (STScI) and NASA: p. 22. NASA/JPL-Caltech/STScI/CXC/SAO/O. Krause (Steward Observatory): p. 23 top. NASA/JPL-Caltech/R. Hurt (SSC-Caltech): p. 23 bottom. Carleton Bailie/United Launch Alliance Feb. 6, 2009: p. 24 top. NASA/JPL/CALTECH/UNIVERSITY OF ARIZONA: p. 24 bottom. NASA, ESA, CXC, C. Ma, H. Ebeling, and E. Barrett (University of Hawaii/IfA), et al., and STScI: p. 26. K. Olsen: p. 27 top. P. Norman: p. 27 bottom. Courtesy of NOAO/AURA/NSF: p. 27 (logo). STScI/NASA: p. 29 top left. NASA, ESA, L. Bradley (Johns Hopkins University), R. Bouwens (University of California, Santa Cruz), H. Ford (Johns Hopkins University), and G. Illingworth (University of California, Santa Cruz): p. 29 top right. E. Acosta: p. 29 bottom. © Iris Nieves: p. 30.